Chapter 4
2-Digit Addition

Houghton
Mifflin
Harcourt

W9-AHF-186

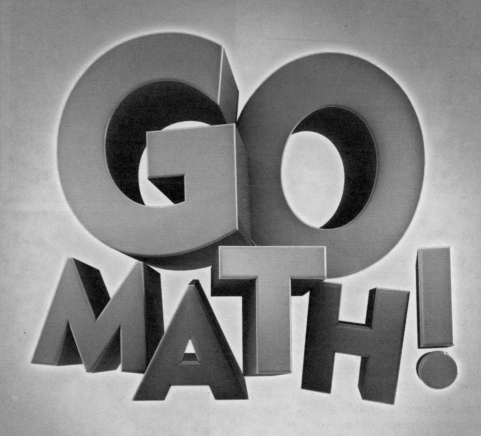

Made in the United States
Text printed on 100%
recycled paper

Houghton
Mifflin
Harcourt

Printed in the U.S.A.

ISBN 978-0-544-34200-2

19 0928 20

4500800118   E F G

Dear Students and Families,

Welcome to **Go Math!**, Grade 2! In this exciting mathematics program, there are hands-on activities to do and real-world problems to solve. Best of all, you will write your ideas and answers right in your book. In **Go Math!**, writing and drawing on the pages helps you think deeply about what you are learning, and you will really understand math!

By the way, all of the pages in your **Go Math!** book are made using recycled paper. We wanted you to know that you can Go Green with **Go Math!**

Sincerely,

The Authors

GO MATH!

Authors

Juli K. Dixon, Ph.D.
Professor, Mathematics Education
University of Central Florida
Orlando, Florida

Edward B. Burger, Ph.D.
President, Southwestern University
Georgetown, Texas

Steven J. Leinwand
Principal Research Analyst
American Institutes for
    Research (AIR)
Washington, D.C.

Contributor

Rena Petrello
Professor, Mathematics
Moorpark College
Moorpark, California

Matthew R. Larson, Ph.D.
K-12 Curriculum Specialist for
    Mathematics
Lincoln Public Schools
Lincoln, Nebraska

Martha E. Sandoval-Martinez
Math Instructor
El Camino College
Torrance, California

English Language Learners Consultant

Elizabeth Jiménez
CEO, GEMAS Consulting
Professional Expert on English
    Learner Education
Bilingual Education and
    Dual Language
Pomona, California

Addition and Subtraction

Critical Area Building fluency with addition and subtraction

GO DIGITAL

Go online! Your math lessons are interactive. Use *iTools*, Animated Math Models, the Multimedia *e*Glossary, and more.

Essential Question
How does breaking apart a number make it easier to add?

Chapter 4 Overview

In this chapter, you will explore and discover answers to the following **Essential Questions**:

- How do you use place value to add 2-digit numbers, and what are some different ways to add 2-digit numbers?
- How do you make an addend a ten to help solve an addition problem?
- How do you record the steps when adding 2-digit numbers?
- What are some ways to add 3 numbers or 4 numbers?

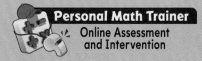

Personal Math Trainer
Online Assessment and Intervention

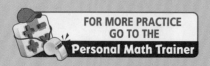

FOR MORE PRACTICE
GO TO THE
Personal Math Trainer

Practice and Homework

Lesson Check and
Spiral Review in
every lesson

2-Digit Addition

Curious about Math

The keys of a modern piano are made from wood or plastic. A modern piano has 36 black keys and 52 white keys. How many keys is this in all?

✓ Show What You Know

Addition Patterns

Add 2. Complete each addition sentence. (1.OA.A.1)

1. $1 + \underline{2} = \underline{3}$

2. $2 + \underline{\hphantom{0}} = \underline{\hphantom{0}}$

3. $3 + \underline{\hphantom{0}} = \underline{\hphantom{0}}$

4. $4 + \underline{\hphantom{0}} = \underline{\hphantom{0}}$

5. $5 + \underline{\hphantom{0}} = \underline{\hphantom{0}}$

6. $6 + \underline{\hphantom{0}} = \underline{\hphantom{0}}$

Addition Facts

Write the sum. (1.OA.C.6)

7. $\begin{array}{r} 7 \\ + 3 \\ \hline \end{array}$
8. $\begin{array}{r} 8 \\ + 8 \\ \hline \end{array}$
9. $\begin{array}{r} 6 \\ + 7 \\ \hline \end{array}$
10. $\begin{array}{r} 4 \\ + 4 \\ \hline \end{array}$
11. $\begin{array}{r} 9 \\ + 5 \\ \hline \end{array}$
12. $\begin{array}{r} 8 \\ + 7 \\ \hline \end{array}$

Tens and Ones

Write how many tens and ones for each number. (1.NBT.B.2b)

13. 43

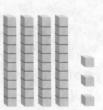

_____ tens _____ ones

14. 68

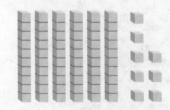

_____ tens _____ ones

This page checks understanding of important skills needed for success in Chapter 4.

Vocabulary Builder

Review Words

sum
addend
digit
tens
ones

Visualize It

Use review words to fill in the graphic organizer.

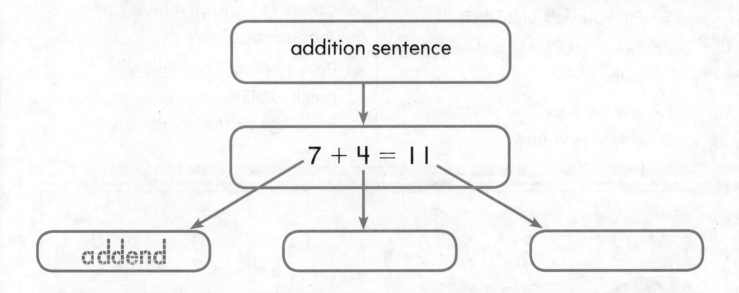

addition sentence

7 + 4 = 11

addend

Understand Vocabulary

1. Write a number with the **digit** 3 in the **tens** place.   _____

2. Write a number with the **digit** 5 in the **ones** place.   _____

3. Write a number that has the same **digit** in the **tens** place and in the **ones** place.   _____

4. Write a number with **digits** that have a **sum** of 8.   _____

GO DIGITAL • Interactive Student Edition
• Multimedia eGlossary

Game What is the Sum?

Materials

- 12
- 12 ⚪
- 1 🎲

Play with a partner.

1. Put your ⚫ on START.
2. Toss the 🎲. Move that many spaces.
3. Say the sum. Your partner checks your answer.

4. If your answer is correct, find that number in the middle of the board. Put one of your ⚫ on that number.
5. Take turns until both players reach FINISH. The player with more ⚫ on the board wins.

START

FINISH

$$\begin{array}{c} 2 \\ +7 \end{array}$$

$$\begin{array}{c} 6 \\ +5 \end{array}$$

$$\begin{array}{c} 3 \\ +9 \end{array}$$

$$\begin{array}{c} 0 \\ +7 \end{array}$$

$$\begin{array}{c} 8 \\ +6 \end{array}$$

$$\begin{array}{c} 9 \\ +8 \end{array}$$

7	18	9	11	15
13	6	17	8	10
16	4	12	14	5

$$\begin{array}{c} 6 \\ +2 \end{array}$$

$$\begin{array}{c} 1 \\ +4 \end{array}$$

$$\begin{array}{c} 8 \\ +7 \end{array}$$

$$\begin{array}{c} 5 \\ +8 \end{array}$$

$$\begin{array}{c} 9 \\ +9 \end{array}$$

$$\begin{array}{c} 7 \\ +9 \end{array}$$

$$\begin{array}{c} 2 \\ +2 \end{array}$$

$$\begin{array}{c} 4 \\ +6 \end{array}$$

$$\begin{array}{c} 5 \\ +1 \end{array}$$

Chapter 4 Vocabulary

column

columna

7

digit

dígito

15

hundred

centena

31

is equal to (=)

es igual a

33

ones

unidades

45

regroup

reagrupar

56

sum

suma o total

59

ten

decena

61

0, 1, 2, 3, 4, 5, 6, 7, 8, and 9 are **digits**.

column

$$\begin{array}{r} 3\,\boxed{3} \\ 3\,4 \\ +\ 3\,2 \end{array}$$

2   plus   1   is equal to   3

2   +   1   =   3

There are 10 tens in **1 hundred**.

Tens	Ones

You can trade 10 ones for 1 ten to **regroup**.

10 ones = 1 ten

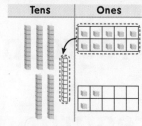

10 ones = 1 ten

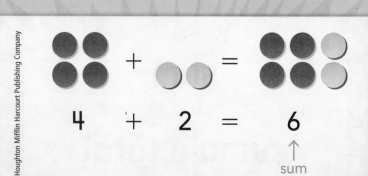

4   +   2   =   6

↑
sum

Matchup

For 2 to 3 players

Materials
- 1 set of word cards

How to Play

1. Put the cards face-down in rows. Take turns to play.
2. Choose two cards. Turn the cards face-up.
   - If the cards match, keep the pair and take another turn.
   - If the cards do not match, turn them over again.
3. The game is over when all cards have been matched. The players count their pairs. The player with the most pairs wins.

Word Box

column
digit
hundred
is equal to (=)
ones
regroup
sum
tens

The Write Way

Reflect

Choose one idea. Write about it in the space below.

- Explain how drawing quick pictures helps you add 2-digit numbers.
- Tell about all the different ways you can add 2-digit numbers.
- Write three things you know about regrouping.

Name _____

Break Apart Ones to Add

Essential Question How does breaking apart a number make it easier to add?

Common Core Number and Operations in Base Ten—2.NBT.B.5
MATHEMATICAL PRACTICES
MP1, MP4, MP6

Listen and Draw Real World  Hands On

Use ▭▭▭ ▪. Draw to show what you did.

FOR THE TEACHER • Read the following problem. Have children use blocks to solve. Griffin read 27 books about animals and 6 books about space. How many books did he read?

Math Talk MATHEMATICAL PRACTICES 6

Describe what you did with the blocks.

Model and Draw

Break apart ones to make a ten.
Use this as a way to add.

27 + 8 = _____?_____

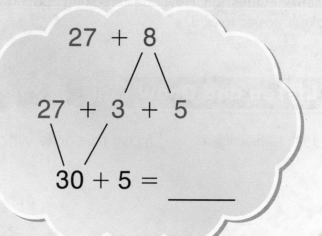

27 + 8

27 + 3 + 5

30 + 5 = _____

27 + 8 = _____

Share and Show MATH BOARD

Draw quick pictures. Break apart ones to
make a ten. Then add and write the sum.

1. 15 + 7 = _____

2. 26 + 5 = _____

3. 37 + 8 = _____

4. 28 + 6 = _____

Name _____

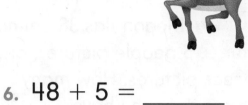

Break apart ones to make a ten.
Then add and write the sum.

5. 23 + 9 = _____

6. 48 + 5 = _____

7. 18 + 5 = _____

8. 33 + 9 = _____

9. 27 + 6 = _____

10. 49 + 4 = _____

11. **GO DEEPER** Alex sets up 32 small tables and 9 large tables in a room. Then he sets up 5 more large tables along a wall. How many tables does Alex set up?

_____ tables

12. **THINK SMARTER** Bruce sees 29 oak trees and 4 maple trees at the park. Then he sees double the number of pine trees as maple trees. How many trees does Bruce see?

_____ trees

Problem Solving • Applications   WRITE Math

Solve. Write or draw to explain.

13. **GO DEEPER** Megan has 38 animal pictures, 5 people pictures, and 3 insect pictures. How many pictures does she have?

_____ pictures

14. **MATHEMATICAL PRACTICE ① Analyze**
Jamal has a box with 22 toy cars in it. He puts 9 more toy cars into the box. Then he takes 3 toy cars out of the box. How many toy cars are in the box now?

_____ toy cars

15. **THINK SMARTER** Dan has 16 pencils. Quentin gives him 5 more pencils. Choose all the ways you can use to find how many pencils Dan has in all.

   ○ $16 + 5$

   ○ $16 + 4 + 1$

   ○ $16 - 5$

TAKE HOME ACTIVITY • Say a number from 0 to 9. Have your child name a number to add to yours to have a sum of 10.

Break Apart Ones to Add

COMMON CORE STANDARD—2.NBT.B.5
Use place value understanding and properties of
operations to add and subtract.

Break apart ones to make a ten.
Then add and write the sum.

1. $62 + 9 =$ _____

2. $27 + 7 =$ _____

3. $28 + 5 =$ _____

4. $17 + 8 =$ _____

5. $57 + 6 =$ _____

6. $23 + 9 =$ _____

7. $39 + 7 =$ _____

8. $26 + 5 =$ _____

9. $13 + 8 =$ _____

10. $18 + 7 =$ _____

Problem Solving

Solve. Write or draw to explain.

11. Jimmy had 18 toy airplanes. His mother
bought him 7 more toy airplanes. How many
toy airplanes does he have now?

_____ toy airplanes

12. **WRITE** **Math** Explain how you would find the sum of $46 + 7$.

_____

_____

Lesson Check (2.NBT.B.5)

1. What is the sum?

$$26 + 7 = \underline{\quad}$$

2. What is the sum?

$$15 + 8 = \underline{\quad}$$

Spiral Review (2.OA.A.1, 2.OA.B.2, 2.NBT.A.3)

3. Hannah has 4 blue beads and 8 red beads. How many beads does Hannah have?

$$4 + 8 = \underline{\quad} \text{ beads}$$

4. Rick had 4 stickers. Then he earned 2 more. How many stickers does he have now?

$$4 + 2 = \underline{\quad} \text{ stickers}$$

5. What is the sum?

$$4 + 5 + 4 = \underline{\quad}$$

6. Write 281 using hundreds, tens, and ones.

$$\underline{\quad} \text{ hundreds} \underline{\quad} \text{ tens} \underline{\quad} \text{ one}$$

FOR MORE PRACTICE GO TO THE Personal Math Trainer

Name _____

Use Compensation

Essential Question How can you make an addend a ten to help solve an addition problem?

Common Core **Number and Operations in Base Ten—2.NBT.B.5**

MATHEMATICAL PRACTICES
MP1, MP4, MP6

Listen and Draw Real World

Draw quick pictures to show the problems.

FOR THE TEACHER • Have children draw quick pictures to solve this problem. Kara has 47 stickers. She buys 20 more stickers. How many stickers does she have now? Repeat for this problem. Tyrone has 30 stickers and buys 52 more stickers. How many stickers does he have now?

 Math Talk

MATHEMATICAL PRACTICES

Analyze Describe how you found how many stickers Tyrone has.

Model and Draw

Take ones from an addend to make the other addend the next tens number.

> Adding can be easier when one of the addends is a tens number.

$$25 + 48 = ?$$

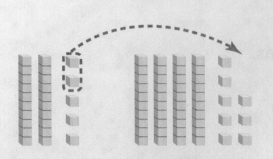

$$\underline{23} + \underline{50} = \underline{}$$

Share and Show MATH BOARD

Show how to make one addend the next tens number.
Complete the new addition sentence.

1. $37 + 25 = ?$

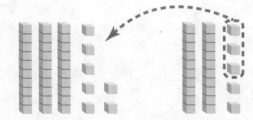

$$\underline{40} + \underline{} = \underline{}$$

2. $27 + 46 = ?$

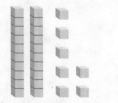 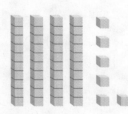

$$\underline{} + \underline{} = \underline{}$$

3. $14 + 29 = ?$

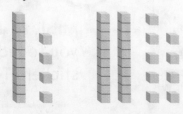

$$\underline{} + \underline{} = \underline{}$$

244 two hundred forty-four

On Your Own

Show how to make one addend the next tens number.
Complete the new addition sentence.

4. $18 + 13 = ?$

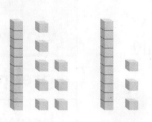

_____ + _____ = _____

5. $24 + 18 = ?$

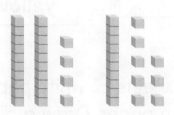

_____ + _____ = _____

6. **GO DEEPER** Lee finds 44 shells. Wayne
finds 39 shells. How many shells do
they still need if they want 90 shells
in all?

_____ shells

Solve. Write or draw to explain.

7. **THINK SMARTER** Zach finds
38 sticks. Kelly finds 27 sticks.
How many more sticks do the
two children still need if they
want 70 sticks in all?

_____ more sticks

Problem Solving • Applications

Solve. Write or draw to explain.

8. **MATHEMATICAL PRACTICE 6** **Make Connections**
The chart shows the leaves that
Philip collected. He wants a
collection of 52 leaves, using only
two colors. Which two colors of
leaves should he use?

Leaves Collected	
Color	**Number**
green	27
brown	29
yellow	25

_____ and _____

9. **THINK SMARTER** Ava has 39 sheets of white paper.
She has 22 sheets of green paper. Draw a picture
and write to explain how to find the number
of sheets of paper Ava has.

Ava has _____ sheets of paper.

 TAKE HOME ACTIVITY • Have your child choose one problem
on this page and explain how to solve it in another way.

Use Compensation

COMMON CORE STANDARD—2.NBT.B.5
Use place value understanding and properties of operations to add and subtract.

Show how to make one addend the next tens number. Complete the new addition sentence.

1. $15 + 37 = ?$

 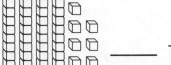

___ + ___ = ___

2. $22 + 49 = ?$

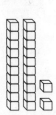

___ + ___ = ___

3. $38 + 26 = ?$

___ + ___ = ___

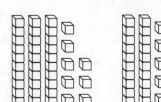

Problem Solving Real World

Solve. Write or draw to explain.

4. The oak tree at the school was 34 feet tall.
   Then it grew 18 feet taller.
   How tall is the oak tree now?

_____ feet tall

5. **WRITE** Math Explain why you would make one of the addends a tens number when solving an addition problem.

_____

_____

Lesson Check (2.NBT.B.5)

1. What is the sum?

$$18 + 25 = \underline{\qquad}$$

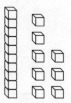 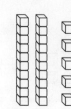

2. What is the sum?

$$27 + 24 = \underline{\qquad}$$

 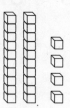

Spiral Review (2.OA.B.2, 2.OA.C.3)

3. Circle the even number.

27   14   11   5

4. Andrew sees 4 fish. Kim sees double that number of fish. How many fish does Kim see?

____ fish

5. Write a related subtraction fact for $7 + 6 = 13$.

_____

6. What is the sum?

$$2 + 8 = \underline{\qquad}$$

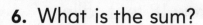

FOR MORE PRACTICE
GO TO THE
Personal Math Trainer

Name _____

Break Apart Addends as Tens and Ones

Essential Question How do you break apart addends to add tens and then add ones?

Common Core
Number and Operations in Base Ten—2.NBT.B.5
MATHEMATICAL PRACTICES
MP1, MP6, MP8

Listen and Draw

Write the number. Then write the number as tens plus ones.

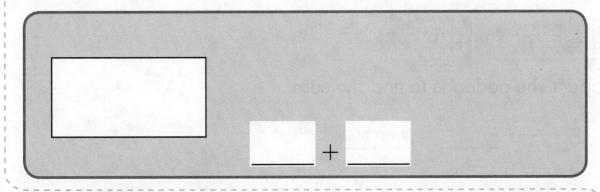

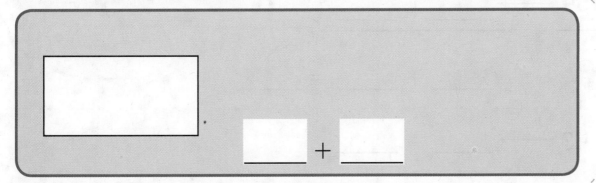

FOR THE TEACHER • Direct children's attention to the orange box. Have children write 25 inside the large rectangle. Then ask children to write 25 as tens plus ones. Repeat the activity for 36 and 42.

Math Talk

MATHEMATICAL PRACTICES 1

What is the value of the 6 in the number 63? **Explain** how you know.

Model and Draw

Break apart the addends into tens and ones.
Add the tens and add the ones.
Then find the total sum.

27 ⟶ 20 + 7
+48 ⟶ 40 + 8

___60___ + ___15___ = _____

60 + 15

10   5

70 + 5 = _____

Share and Show MATH BOARD

Break apart the addends to find the sum.

1.   35 ⟶ _____ + _____

     +54 ⟶ _____ + _____
     _____

         _____ + _____ = _____

✓ 2.   43 ⟶ _____ + _____

      +29 ⟶ _____ + _____
      _____

          _____ + _____ = _____

✓ 3.   56 ⟶ _____ + _____

      +38 ⟶ _____ + _____
      _____

          _____ + _____ = _____

250 two hundred fifty

Name _____

Break apart the addends to find the sum.

4.   14 ⟶ _____ + _____

  +23 ⟶ _____ + _____
  _____

  _____ + _____ = _____

5.   37 ⟶ _____ + _____

  +45 ⟶ _____ + _____
  _____

  _____ + _____ = _____

6. **GO DEEPER** Chris read 15 pages of his book. Tony read 4 more pages than Chris. How many pages did Chris and Tony read?

_____ pages

7. **THINK SMARTER** Julie read 18 pages of her book in the morning. She read the same number of pages in the afternoon. How many pages did she read?

_____ pages

Problem Solving • Applications    WRITE Math

Write or draw to explain.

8. **MATHEMATICAL PRACTICE ❶ Make Sense of Problems** Len has 35 baseball cards. The rest of his cards are basketball cards. He has 58 cards in all. How many basketball cards does he have?

_____ basketball cards

9. **MATHEMATICAL PRACTICE ❶ Evaluate** Tomás has 17 pencils. He buys 26 more pencils. How many pencils does Tomás have now?

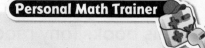

_____ pencils

Personal Math Trainer

10. **THINK SMARTER +** Sasha used 38 red stickers and 22 blue stickers. Show how you can break apart the addends to find how many stickers Sasha used.

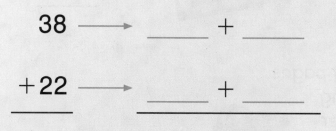

$$38 \longrightarrow \underline{\hspace{1cm}} + \underline{\hspace{1cm}}$$

$$+22 \longrightarrow \underline{\hspace{1cm}} + \underline{\hspace{1cm}}$$

$$\underline{\hspace{1cm}} + \underline{\hspace{1cm}} = \underline{\hspace{1cm}} \text{ stickers}$$

 TAKE HOME ACTIVITY • Write 32 + 48 on a sheet of paper. Have your child break apart the numbers and find the sum.

Break Apart Addends as Tens and Ones

Common Core **COMMON CORE STANDARD—2.NBT.B.5**
Use place value understanding and properties of operations to add and subtract.

Break apart the addends to find the sum.

1.   18 → ___ + ___
    + 21 → ___ + ___
             ___ + ___ = ___

2.   33 → ___ + ___
    + 49 → ___ + ___
             ___ + ___ = ___

Problem Solving

Choose a way to solve.
Write or draw to explain.

3. Christopher has 28 baseball cards.
   Justin has 18 baseball cards. How
   many baseball cards do they
   have together?   ____ baseball cards

4. **WRITE** Math Explain how to break apart the addends
   to find the sum of 25 + 16.

   _____

   _____

Lesson Check (2.NBT.B.5)

1. What is the sum?

$$\begin{array}{r} 27 \\ + 12 \\ \hline \end{array}$$

2. What is the sum?

$$\begin{array}{r} 17 \\ + 35 \\ \hline \end{array}$$

Spiral Review (2.OA.B.2, 2.NBT.A.1, 2.NBT.A.3, 2.NBT.B.5)

3. What is the value of the underlined digit?

2<u>5</u>

_____

4. What number has the same value as 12 tens?

_____

5. Ally has 7 connecting cubes. Greg has 4 connecting cubes. How many connecting cubes do they have?

_____ cubes

6. Juan painted a picture of a tree. First he painted 15 leaves. Then he painted 23 more leaves. How many leaves did he paint?

_____ leaves

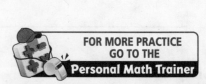

FOR MORE PRACTICE
GO TO THE
Personal Math Trainer

Name _____

Model Regrouping for Addition

Essential Question When do you regroup in addition?

Common Core **Number and Operations in Base Ten—2.NBT.B.5**
MATHEMATICAL PRACTICES
MP1, MP5, MP7

Listen and Draw *Real World* Hands On

Use ▭▭▭▭ ▫ to model the problem.
Draw quick pictures to show what you did.

Tens	Ones

 Math Talk

MATHEMATICAL PRACTICES 5

Use Tools Describe how you made a ten in your model.

 FOR THE TEACHER • Read the following problem. Brandon has 24 books. His friend Mario has 8 books. How many books do they have?

Model and Draw

Add 37 and 25.

Step 1 Look at the ones. Can you make a ten?

Tens	Ones

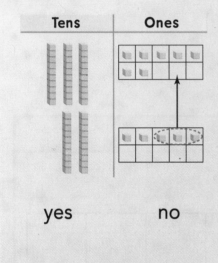

yes   no

Step 2 If you can make a ten, **regroup**.

Tens	Ones

Trade 10 ones for 1 ten to regroup.

Step 3 Write how many tens and ones. Write the sum.

Tens	Ones

_____ tens _____ ones

_____

Share and Show MATH BOARD

Draw to show the regrouping. Write how many tens and ones are in the sum. Write the sum.

1. Add 47 and 15.

Tens	Ones

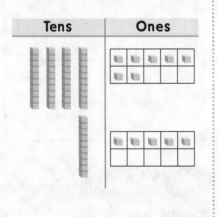

_____ tens _____ ones

_____

2. Add 48 and 8.

Tens	Ones

_____ tens _____ ones

_____

3. Add 26 and 38.

Tens	Ones

_____ tens _____ ones

_____

Name _____

On Your Own

Draw to show if you regroup. Write how many
tens and ones are in the sum. Write the sum.

4. Add 79 and 6.

Tens	Ones

_____ tens _____ ones

_____

5. Add 18 and 64.

Tens	Ones

_____ tens _____ ones

_____

6. Add 23 and 39.

Tens	Ones

_____ tens _____ ones

_____

7. Add 54 and 25.

Tens	Ones

_____ tens _____ ones

_____

8. Add 33 and 7.

Tens	Ones

_____ tens _____ ones

_____

9. Add 27 and 68.

Tens	Ones

_____ tens _____ ones

_____

10. **THINK SMARTER** Kara has 25 toy animals
and 12 books. Jorge has 8 more toy
animals than Kara has. How many
toy animals does Jorge have?

_____ toy animals

Problem Solving • Applications

WRITE Math

Write or draw to explain.

11. **MATHEMATICAL PRACTICE ①** Make Sense of Problems Mrs. Sanders has two fish tanks. There are 14 fish in the small tank. There are 27 fish in the large tank. How many fish are in the two tanks?

_____ fish

12. **THINK SMARTER** Charlie climbed 69 steps. Then he climbed 18 more steps. Show two different ways to find how many steps Charlie climbed.

Charlie climbed _____ steps.

TAKE HOME ACTIVITY • Ask your child to write a word problem with 2-digit numbers about adding two groups of stamps.

Name _____

Model Regrouping for Addition

Draw to show the regrouping. Write how many tens and ones in the sum. Write the sum.

Common Core **COMMON CORE STANDARD—2.NBT.B.5**
Use place value understanding and properties of operations to add and subtract.

1. Add 63 and 9.

Tens	Ones

_____ tens _____ ones

_____

2. Add 25 and 58.

Tens	Ones

_____ tens _____ ones

_____

3. Add 58 and 18.

Tens	Ones

_____ tens _____ ones

_____

4. Add 64 and 26.

Tens	Ones

_____ tens _____ ones

_____

5. Add 17 and 77.

Tens	Ones

_____ tens _____ ones

_____

6. Add 16 and 39.

Tens	Ones

_____ tens _____ ones

_____

Problem Solving

Choose a way to solve. Write or draw to explain.

7. Cathy has 43 leaves in her collection. Jane has 38 leaves. How many leaves do the two children have?

_____ leaves

8. **WRITE** Math Suppose you are adding 43 and 28. Will you regroup? Explain.

_____

Lesson Check (2.NBT.B.5)

1. Add 27 and 48. What is the sum?

Tens	Ones

_____

Spiral Review (2.OA.B.2, 2.OA.C.3, 2.NBT.B.5)

2. What is the sum?

$7 + 7 =$ _____

3. Circle the odd number.

6   12   21   22

4. What is the sum?

$39 + 46 =$ _____

5. What is the sum?

$5 + 3 + 4 =$ _____

© Houghton Mifflin Harcourt Publishing Company

FOR MORE PRACTICE
GO TO THE
Personal Math Trainer

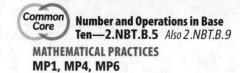
Model and Record 2-Digit Addition

Essential Question How do you record 2-digit addition?

Common Core Number and Operations in Base Ten—2.NBT.B.5 *Also 2.NBT.B.9*
MATHEMATICAL PRACTICES
MP1, MP4, MP6

Listen and Draw

Use ▭▭▭▭ ▪ to model the problem.
Draw quick pictures to show what you did.

Tens	Ones

Math Talk

MATHEMATICAL PRACTICES 6

Make Connections Did you trade blocks in your model? Explain why or why not.

FOR THE TEACHER • Read the following problem. Mr. Riley's class collected 54 cans for the food drive. Miss Bright's class collected 35 cans. How many cans did the two classes collect?

Model and Draw

Trace over the quick pictures in the steps.

Step 1 Model 37 + 26. Are there 10 ones to regroup?

Tens	Ones
3	7
+ 2	6

Step 2 Write the regrouped ten. Write how many ones are in the ones place now.

Tens	Ones
1	
3	7
+ 2	6
	3

Step 3 How many tens are there? Write how many tens are in the tens place.

Tens	Ones
1	
3	7
+ 2	6
6	3

Share and Show MATH BOARD

Draw quick pictures to help you solve. Write the sum.

✓ 1.

Tens	Ones
2	6
+ 3	2

Tens	Ones

✓ 2.

Tens	Ones
5	8
+ 2	4

Tens	Ones

Name _____

Draw quick pictures to help you solve. Write the sum.

3.

Tens	Ones
☐	
3	4
+	9

Tens	Ones

4.

Tens	Ones
☐	
2	7
+ 2	4

Tens	Ones

5.

Tens	Ones
☐	
3	5
+ 2	3

Tens	Ones

6.

Tens	Ones
☐	
5	9
+	6

Tens	Ones

7. **THINK SMARTER** Tim has 36 stickers. Margo has 44 stickers. How many more stickers would they need to have 100 stickers altogether?

_____ more stickers

8. **GO DEEPER** A baker wants to sell 100 muffins. So far the baker has sold 48 corn muffins and 42 bran muffins. How many more muffins does the baker need to sell?

_____ more muffins

Problem Solving • Applications

 **Math**

Write or draw to explain.

9. **MATHEMATICAL PRACTICE ①** **Make Sense of Problems**
Chris and Bianca got 80 points in all in the spelling contest. Each child got more than 20 points. How many points could each child have gotten?

Chris: _____ points

Bianca: _____ points

Personal Math Trainer

10. **THINK SMARTER +** Don built a tower with 24 blocks. He built another tower with 18 blocks. How many blocks did Don use for both towers? Draw quick pictures to solve. Write the sum.

Tens	Ones

_____ blocks

Did you regroup to find the answer? Explain.

_____

_____

_____

 TAKE HOME ACTIVITY • Write two 2-digit numbers and ask your child if he or she would regroup to find the sum.

Name _____

Model and Record 2-Digit Addition

Common Core **COMMON CORE STANDARD—2.NBT.B.5**
Use place value understanding and properties of operations to add and subtract.

**Draw quick pictures to help you solve.
Write the sum.**

1.

Tens	Ones
□	
3	8
+ 1	7

Tens	Ones

2.

Tens	Ones
□	
5	8
+ 2	6

Tens	Ones

3.

Tens	Ones
□	
4	2
+ 3	7

Tens	Ones

4.

Tens	Ones
□	
5	3
+ 3	8

Tens	Ones

Problem Solving

Choose a way to solve.
Write or draw to explain.

5. There were 37 children at the park on
Saturday and 25 children at the park
on Sunday. How many children were
at the park on those two days?

_____ children

6. **WRITE** Math Explain why you should record a 1 in the
Tens column when you regroup in an addition problem.

_____

_____

Lesson Check (2.NBT.B.5)

1. What is the sum?

Tens	Ones
□	
3	4
+ 2	8

2. What is the sum?

Tens	Ones
□	
4	3
+ 2	7

Spiral Review (2.OA.B.2)

3. Adam collected 14 pennies in the first week and 9 pennies in the second week. How many more pennies did he collect in the first week than in the second week?

14 − 9 = _____ pennies

4. What is the sum?

3 + 7 + 9 = _____

5. Janet has 5 marbles. She finds double that number of marbles in her art box. How many marbles does Janet have now?

5 + _____ = _____ marbles

6. What is the difference?

13 − 5 = _____

FOR MORE PRACTICE
GO TO THE
Personal Math Trainer

Name _____

2-Digit Addition

Essential Question How do you record the steps when adding 2-digit numbers?

Common Core **Number and Operations in Base Ten—2.NBT.B.5, 2.NBT.B.9**
MATHEMATICAL PRACTICES
MP1, MP3, MP6

 **Listen and Draw** Real World

Draw quick pictures to model each problem.

Tens	Ones

Tens	Ones

FOR THE TEACHER • Read the following problem and have children draw quick pictures to solve. Jason scored 35 points in one game and 47 points in another game. How many points did Jason score? Repeat the activity with this problem. Patty scored 18 points. Then she scored 21 points. How many points did she score in all?

Math Talk
MATHEMATICAL PRACTICES

Analyze Relationships
Explain why regrouping works.

© Houghton Mifflin Harcourt Publishing Company

Model and Draw

Add 59 and 24.

Step 1 Add the ones.	Step 2 Regroup.	Step 3 Add the tens.
$9 + 4 = 13$	13 ones is the same as 1 ten 3 ones.	$1 + 5 + 2 = \mathbf{8}$

Tens	Ones

Tens	Ones
☐	
5	9
+ 2	4

Tens	Ones
1	
5	9
+ 2	4
	3

Tens	Ones
1	
5	9
+ 2	4
8	3

Share and Show MATH BOARD

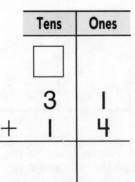

Regroup if you need to. Write the sum.

1.

Tens	Ones
☐	
4	2
+ 2	9

✓2.

Tens	Ones
☐	
3	1
+ 1	4

✓3.

Tens	Ones
☐	
2	7
+ 4	5

Name _____

On Your Own

Regroup if you need to. Write the sum.

4.

Tens	Ones
☐	
4	8
+	7

5.

Tens	Ones
☐	
3	5
+ 4	2

6.

Tens	Ones
☐	
7	3
+ 2	0

7.

3	3
+ 2	7

8.

5	2
+	5

9.

3	6
+ 5	8

10.

6	4
+ 2	5

11.

3	5
+ 3	8

12.

3	8
+ 5	2

Solve. Write or draw to explain.

13. THINK SMARTER Jin has 31 books about cats and 19 books about dogs. He gives 5 books to his sister. How many books does Jin have now?

_____ books

Problem Solving • Applications   WRITE Math

14. GO DEEPER Abby used a different way to add. Find the sum using Abby's way.

$$
\begin{array}{r}
35 \\
+\ 48 \\
\hline
13 \\
+\ 70 \\
\hline
83
\end{array}
$$

$$
\begin{array}{r}
5\ 7 \\
+\ 2\ 9 \\
\hline

\end{array}
$$

15. MATHEMATICAL PRACTICE ③ **Verify the Reasoning of Others**
Explain why Abby's way works.

_____

_____

_____

16. THINK SMARTER Melissa saw 14 sea lions and 29 seals. How many animals did she see? Write a number sentence to find the total number of animals that she saw.

_____

Explain how the number sentence shows the problem.

_____

_____

 TAKE HOME ACTIVITY • Ask your child to show you two ways to add 45 and 38.

2-Digit Addition

 COMMON CORE STANDARD—2.NBT.B.5
Use place value understanding and properties of operations to add and subtract.

Regroup if you need to. Write the sum.

1.
```
   4 | 7
 + 2 | 5
```

2.
```
   3 | 3
 + 1 | 8
```

3.
```
   2 | 8
 + 6 | 4
```

4.
```
   1 | 3
 + 6 | 5
```

5.
```
   1 | 7
 + 2 | 6
```

6.
```
   3 | 6
 + 5 | 3
```

7.
```
   5 | 8
 + 2 | 5
```

8.
```
   3 | 7
 + 4 | 9
```

 Problem Solving

Solve. Write or draw to explain.

9. Angela drew 16 flowers on her paper in the morning. She drew 25 more flowers in the afternoon. How many flowers did she draw?

_____ flowers

10. **Math** How is Exercise 5 different from Exercise 6? Explain.

_____

_____

Lesson Check (2.NBT.B.5)

1. What is the sum?

$$\begin{array}{r} 2\,|\,1 \\ +\ 3\,|\,7 \\ \hline \end{array}$$

2. What is the sum?

$$\begin{array}{r} 3\,|\,8 \\ +\ 5\,|\,2 \\ \hline \end{array}$$

Spiral Review (2.OA.A.1, 2.NBT.A.3, 2.NBT.B.8)

3. What is the next number in the counting pattern?

103, 203, 303, 403, ____

4. Rita counted 13 bubbles. Ben counted 5 bubbles. How many fewer bubbles did Ben count than Rita?

$13 - 5 =$ ____ bubbles

5. Which number is 100 more than 265?

____

6. Write 42 as a sum of tens and ones.

____ + ____

FOR MORE PRACTICE
GO TO THE
Personal Math Trainer

Practice 2-Digit Addition

Essential Question How do you record the steps when adding 2-digit numbers?

Common Core Number and Operations in Base Ten—2.NBT.B.5 *Also 2.NBT.B.7*

MATHEMATICAL PRACTICES
MP1, MP6, MP7

Listen and Draw · Real World

Choose one way to solve the problem. Draw or write to show what you did.

 FOR THE TEACHER • Read the following problem. There were 45 boys and 53 girls who ran in the race. How many children ran in the race?

Math Talk

 MATHEMATICAL PRACTICES 6

Explain why you chose your way of solving the problem.

Model and Draw

Mrs. Meyers sold 47 snacks before the game. Then she sold 85 snacks during the game. How many snacks did she sell?

Step 1 Add the ones.

$7 + 5 = 12$

Regroup 12 ones as 1 ten 2 ones.

```
  1
  4 7
+ 8 5
  ___
    2
```

Step 2 Add the tens.

$1 + 4 + 8 = 13$

```
  1
  4 7
+ 8 5
  ___
    2
```

Step 3 13 tens can be regrouped as 1 hundred 3 tens. Write the hundreds digit and the tens digit in the sum.

```
  1
  4 7
+ 8 5
  ___
 13 2
```

Share and Show

Write the sum.

1.
```
  3 8
+ 9 4
```

2.
```
  4 5
+ 5 2
```

3.
```
  8 3
+ 7 6
```

4.
```
  5 6
+ 3 5
```

☑5.
```
  6 3
+ 5 1
```

☑6.
```
  7 4
+ 4 9
```

Name _____

On Your Own

Write the sum.

7.

```
  5 2
+ 3 7
```

8.

```
  8 8
+ 2 1
```

9.

```
  7 4
+ 6 7
```

10.

```
  9 3
+ 5 4
```

11.

```
  9 2
+ 7 8
```

12.

```
  5 6
+ 1 6
```

13.

```
  3 1
+ 4 5
```

14.

```
  4 3
+ 7 2
```

15. **THINK SMARTER** Without finding the sums, circle the pairs of addends for which the sum will be greater than 100.

Explain how you decided which pairs to circle.

73
18

54
71

47
62

36
59

_____

_____

_____

_____

TAKE HOME ACTIVITY • Tell your child two 2-digit numbers. Have him or her write the numbers and find the sum.

Name _____

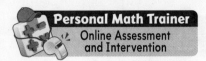
Concepts and Skills

Break apart ones to make a ten.
Then add and write the sum. (2.NBT.B.5)

1. $37 + 8 = $ _____

2. $55 + 7 = $ _____

Break apart the addends to find the sum. (2.NBT.B.5)

3. $27 \longrightarrow$ _____ + _____

   $+36 \longrightarrow$ _____ + _____

   _____ + _____ = _____

Write the sum. (2.NBT.B.5)

4.
$$\begin{array}{r} 2\ 8 \\ +\ 5\ 7 \\ \hline \end{array}$$

5.
$$\begin{array}{r} 6\ 7 \\ +\ 3\ 1 \\ \hline \end{array}$$

6.
$$\begin{array}{r} 7\ 1 \\ +\ 1\ 9 \\ \hline \end{array}$$

7. **THINK SMARTER** Julia collected 25 cans to recycle. Dan collected 14 cans. How many cans did they collect? (2.NBT.B.5)

_____ cans

Practice 2-Digit Addition

Common Core

COMMON CORE STANDARD—2.NBT.B.5
Use place value understanding and properties of operations to add and subtract.

Write the sum.

1.
$$\begin{array}{r} 58 \\ + 17 \\ \hline \end{array}$$

2.
$$\begin{array}{r} 44 \\ + 86 \\ \hline \end{array}$$

3.
$$\begin{array}{r} 36 \\ + 13 \\ \hline \end{array}$$

4.
$$\begin{array}{r} 49 \\ + 72 \\ \hline \end{array}$$

5.
$$\begin{array}{r} 58 \\ + 87 \\ \hline \end{array}$$

6.
$$\begin{array}{r} 32 \\ + 59 \\ \hline \end{array}$$

Problem Solving

Solve. Write or draw to explain.

7. There are 45 books on the shelf.
There are 37 books on the table.
How many books are on the shelf
and the table?

_____ books

8. **WRITE Math** Describe how you regroup when you find the
sum of 64 + 43.

_____

_____

Lesson Check (2.NBT.B.5)

1. What is the sum?

$$\begin{array}{r} 56 \\ + 35 \\ \hline \end{array}$$

2. What is the sum?

$$\begin{array}{r} 74 \\ + 15 \\ \hline \end{array}$$

Spiral Review (2.OA.A.1, 2.OA.B.2, 2.NBT.A.1, 2.NBT.A.3)

3. What is the value of the underlined digit?

5̲26

_____

4. Mr. Stevens wants to put 17 books on the shelf. He put 8 books on the shelf. How many more books does he need to put on the shelf?

$17 - 8 =$ _____ books

5. What is the difference?

$11 - 6 =$ _____

6. Write 83 as a sum of tens and ones.

_____ + _____

FOR MORE PRACTICE GO TO THE
Personal Math Trainer

Name _____

Rewrite 2-Digit Addition

Essential Question What are two different ways to write addition problems?

Common Core
Number and Operations in Base Ten—2.NBT.B.5
MATHEMATICAL PRACTICES
MP1, MP6, MP7

Listen and Draw Real World

Write the numbers for each addition problem.

+ _____

+ _____

+ _____

+ _____

Math Talk MATHEMATICAL PRACTICES 7

Look for Structure Explain why it is important to line up the digits of these addends in columns.

FOR THE TEACHER • Read the following problem and have children write the addends in vertical format. Juan's family drove 32 miles to his grandmother's house. Then they drove 14 miles to his aunt's house. How many miles did they drive? Repeat for three more problems.

Chapter 4

two hundred seventy-nine **279**

Add. 28 + 45 = ?

Step 1 For 28, write the tens digit in the tens column.

Write the ones digit in the ones column.

Repeat for 45.

```
  2 8
+ 4 5
-----
```

Step 2 Add the ones.

Regroup if you need to.

Add the tens.

```
  2 8
+ 4 5
-----
```

Share and Show MATH BOARD

Rewrite the addition problem. Then add.

1. 25 + 8

+ _____

2. 37 + 10

+ _____

3. 25 + 45

+ _____

4. 38 + 29

+ _____

5. 20 + 45

+ _____

6. 63 + 9

+ _____

✓ 7. 15 + 36

+ _____

✓ 8. 74 + 18

+ _____

280 two hundred eighty

Name _____

Rewrite the addition problem. Then add.

9. 27 + 54

10. 34 + 30

11. 26 + 17

12. 48 + 38

+	+	+	+

13. 50 + 32

14. 61 + 38

15. 37 + 43

16. 79 + 17

+	+	+	+

17. 45 + 40

18. 21 + 52

19. 17 + 76

20. 68 + 29

+	+	+	+

21. **THINK SMARTER** For which of the problems above could you find the sum without rewriting it? Explain.

_____

_____

_____

Problem Solving • Applications  Math

Use the table.
Write or draw to
show how you
solved the problem.

Points Scored This Season	
Player	**Number of Points**
Anna	26
Lou	37
Becky	23
Kevin	19

22. **MATHEMATICAL PRACTICE ❶ Analyze Relationships**
Which two players scored 56 points
in all? Add to check your answer.

_____ and_____

23. **THINK SMARTER** Shawn says he can find the sum
of 20 + 63 without rewriting it. Explain how
to find the sum using mental math.

_____

_____

_____

_____

TAKE HOME ACTIVITY • Have your child write and
solve another problem, using the table above.

Rewrite 2-Digit Addition

Common Core

COMMON CORE STANDARD—2.NBT.B.5
Use place value understanding and properties of operations to add and subtract

Rewrite the numbers. Then add.

1. 27 + 19	**2.** 36 + 23	**3.** 31 + 29	**4.** 48 + 23
+ _____	+ _____	+ _____	+ _____
5. 53 + 12	**6.** 69 + 13	**7.** 24 + 38	**8.** 46 + 37
+ _____	+ _____	+ _____	+ _____

Problem Solving Real World

Use the table. Show how you solved the problem.

9. How many pages did Sasha and Kara read together?

_____ pages

Pages Read This Week	
Child	**Number of Pages**
Sasha	62
Kara	29
Juan	50

10. WRITE Math Explain what can happen if you line up the digits incorrectly when you rewrite addition problems.

_____

_____

_____

_____

Lesson Check (2.NBT.B.5)

1. What is the sum of 39 + 17?

$$+ \underline{}$$

2. What is the sum of 28 + 16?

$$+ \underline{}$$

Spiral Review (2.OA.C.4, 2.NBT.A.1, 2.NBT.A.3, 2.NBT.B.6)

3. What number is another way to write 60 + 4?

_____

4. The classroom has 4 desks in each row. There are 5 rows. How many desks are there in the classroom?

_____ desks

5. A squirrel collected 17 acorns. Then the squirrel collected 31 acorns. How many acorns did the squirrel collect?

_____ acorns

6. What number can be written as 3 hundreds 7 tens 5 ones?

© Houghton Mifflin Harcourt Publishing Company

FOR MORE PRACTICE
GO TO THE
Personal Math Trainer

Name _____

Problem Solving • Addition

Essential Question How can drawing a diagram help when solving addition problems?

Common Core
Operations and Algebraic Thinking—2.OA.A.1 Also 2.NBT.B.5
MATHEMATICAL PRACTICES
MP1, MP2, MP4

Kendra had 13 crayons. Her dad gave her some more crayons. Then she had 19 crayons. How many crayons did Kendra's dad give her?

Unlock the Problem

What do I need to find?

how many crayons
_____
Kendra's dad gave her

What information do I need to use?

She had _____ crayons.

After he gave her some more crayons, she had

_____ crayons.

Show how to solve the problem.

There are 19 crayons in all.

13	_____

19

$13 + \blacksquare = 19$ _____

_____ crayons

HOME CONNECTION • Your child used a bar model and a number sentence to represent the problem. These help show what the missing amount is in order to solve the problem.

Label the bar model. Write a number sentence with a ▨ for the missing number. Solve.

• What do I need to find?
• What information do I need to use?

1. Mr. Kane has 24 red pens. He buys 19 blue pens. How many pens does he have now?

_____   _____ pens

2. Hannah has 10 pencils. Jim and Hannah have 17 pencils altogether. How many pencils does Jim have?

_____   _____ pencils

Math Talk

MATHEMATICAL PRACTICES 2

Explain how you know if an amount is a part or the whole in a problem.

Share and Show

Label the bar model. Write a number sentence with a ▮ for the missing number. Solve.

3. Aimee and Matthew catch 17 crickets in all. Aimee catches 9 crickets. How many crickets does Matthew catch?

_____ crickets

_____

4. Percy counts 16 grasshoppers at the park. He counts 15 grasshoppers at home. How many grasshoppers does Percy count?

_____ grasshoppers

_____

5. THINK SMARTER There are three groups of owls. There are 17 owls in each of the first two groups. There are 47 owls in all. How many owls are in the third group?

_____ owls

On Your Own  **WRITE** Math

Write or draw to explain.

6. There are 37 paper clips in the box and 24 paper clips on the table. How many paper clips are there in all?

_____ paper clips

7. **MATHEMATICAL PRACTICE ①** **Make Sense of Problems**
Jeff has 19 postcards and 2 pens. He buys 20 more postcards. How many postcards does he have now?

_____ postcards

8. **GO DEEPER** There are a total of 41 chickens on the farm. There are 13 chickens in each of the 2 cages in the barn. The rest of the chickens are outside. How many chickens are outside?

_____ chickens

9. **THINK SMARTER** There are 23 books in a box. There are 29 books on a shelf. How many books are there?

_____ books

 TAKE HOME ACTIVITY • Ask your child to explain how to solve one of the problems above.

Problem Solving • Addition

COMMON CORE STANDARD—2.OA.A.1
Represent and solve problems involving
addition and subtraction.

Label the bar model. Write a number sentence with a ▇ for the missing number. Solve.

1. Jacob counts 37 ants on the sidewalk and 11 ants on the grass. How many ants does Jacob count?

_____

_____

____ ants

2. There are 14 bees in the hive and 17 bees in the garden. How many bees are there?

_____

_____

____ bees

3. **WRITE Math** Describe how you labeled the bar model and wrote a number sentence to solve Exercise 2.

_____
_____
_____
_____
_____

Lesson Check (2.OA.A.1)

1. Sean and Abby have 23 markers altogether. Abby has 14 markers. How many markers does Sean have?

_____

2. Mrs. James has 22 students in her class. Mr. Williams has 24 students in his class. How many students are in the two classes?

_____

Spiral Review (2.OA.B.2, 2.NBT.A.2)

3. What is the difference?

$$15 - 9 = \underline{\hspace{1cm}}$$

4. What is the sum?

$$7 + 5 = \underline{\hspace{1cm}}$$

5. Jan has 14 blocks. She gives 9 blocks to Tim. How many blocks does Jan have now?

$14 - 9 = \underline{\hspace{1cm}}$ blocks

6. What is the next number in the counting pattern?

29, 39, 49, 59, _____

FOR MORE PRACTICE
GO TO THE
Personal Math Trainer

Algebra • Write Equations to Represent Addition

Essential Question How do you write a number sentence to represent a problem?

Common Core
Operations and Algebraic Thinking—2.OA.A.1 *Also 2.NBT.B.5*
MATHEMATICAL PRACTICES
MP2, MP5, MP6

Listen and Draw Real World

Draw to show how you found the answer.

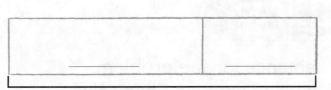

FOR THE TEACHER • Read the following problem and have children choose their own methods for solving. There are 15 children on the bus. Then 9 more children get on the bus. How many children are on the bus now?

Math Talk
MATHEMATICAL PRACTICES 5
Communicate Explain how you found the number of children on the bus.

Model and Draw

You can write a number sentence to show a problem.

Sandy has 16 pencils. Nancy has 13 pencils. How many pencils do the two girls have?

$$16 + 13 = \blacksquare$$

THINK:

$$\begin{array}{r} 16 \text{ pencils} \\ + 13 \text{ pencils} \\ \hline 29 \text{ pencils} \end{array}$$

The two girls have _____ pencils.

Share and Show MATH BOARD

Write a number sentence for the problem. Use a ■ for the missing number. Then solve.

☑1. Carl sees 25 melons at the store. 15 are small and the rest are large. How many melons are large?

_____

_____ melons

☑2. 83 people went to a movie on Thursday. 53 of them were children and the rest were adults. How many adults were at the movie?

_____

_____ adults

Name _____

Write a number sentence for the problem.
Use a ■ for the missing number. Then solve.

3. Jake had some stamps. Then
   he bought 20 more stamps.
   Now he has 56 stamps. How
   many stamps did Jake have
   to start?

_____   _____ stamps

4. [THINK SMARTER] Braden's class went to the
   park. They saw 26 oak trees and
   14 maple trees. They also saw 13 cardinals
   and 35 blue jays. Compare the number of
   trees and the number of birds that
   the class saw.

   _____ ◯ _____

Math on the Spot

5. [MATHEMATICAL PRACTICE 6] Explain Amy needs about
   70 paper clips. Without adding,
   circle 2 boxes that would be close
   to the amount that she needs.

   Explain how you made your choices.

| 70 clips | 81 clips | 54 clips |
| 19 clips | 35 clips | 32 clips |

_____

_____

_____

Problem Solving • Applications

 WRITE Math

6. **MATHEMATICAL PRACTICE ① Make Sense of Problems**

Mr. Walton baked 24 loaves of bread last week. He baked 28 loaves of bread this week. How many loaves of bread did he bake in the two weeks?

_____ loaves of bread

7. **THINK SMARTER** Denise saw these bags of oranges at the store.

Denise bought 26 oranges. Which two bags of oranges did she buy?

Draw or write to show how you solved the problem.

Explain how you found the numbers that have a sum of 26.

_____

_____

_____

TAKE HOME ACTIVITY • Have your child explain how he or she writes a number sentence to stand for a problem.

Algebra • Write Equations to Represent Addition

Common Core

COMMON CORE STANDARD—2.OA.A.1
Represent and solve problems involving addition and subtraction.

Write a number sentence for the problem.
Use a **for the missing number. Then solve.**

1. Emily and her friends went to the park. They saw 15 robins and 9 blue jays. How many birds did they see?

   _____   _____ birds

2. Joe has 13 fish in one tank. He has 8 fish in another tank. How many fish does Joe have?

   _____   _____ fish

Problem Solving

Solve.

3. There are 21 children in Kathleen's class. 12 of the children are girls. How many children in her class are boys?

   _____ boys

4. **WRITE** **Math** Explain how you decided what number sentence to write for Exercise 1.

   _____

   _____

   _____

Lesson Check (2.OA.A.1)

1. Clare has 14 blocks. Jasmine has 6 blocks. How many blocks do they have?

$$14 + 6 = \underline{\quad} \text{ blocks}$$

2. Matt finds 16 acorns at the park. Trevor finds 18 acorns. How many acorns do they find?

$$16 + 18 = \underline{\quad} \text{ acorns}$$

Spiral Review (2.OA.A.1, 2.OA.B.2, 2.OA.C.3, 2.OA.C.4)

3. Leanne counted 19 ants. Gregory counted 6 ants. How many more ants did Leanne count than Gregory?

$$19 - 6 = \underline{\quad} \text{ ants}$$

4. What is the sum?

$$4 + 3 + 6 = \underline{\quad}$$

5. Ms. Santos puts seashells into 4 rows. She puts 6 seashells in each row. How many seashells are there altogether?

$$\underline{\quad} \text{ seashells}$$

6. Circle the even number.

9   14   17   21

FOR MORE PRACTICE
GO TO THE
Personal Math Trainer

Name _____

Algebra • Find Sums for 3 Addends

Essential Question What are some ways to add 3 numbers?

Common Core
Number and Operations in
Base Ten—2.NBT.B.6
MATHEMATICAL PRACTICES
MP1, MP4, MP8

Listen and Draw

Draw to show each problem.

FOR THE TEACHER • Read the following problem and have children draw to show it. Mr. Kim bought 5 blue balloons, 4 red balloons, and 5 yellow balloons. How many balloons did Mr. Kim buy? Repeat for another problem.

Math Talk MATHEMATICAL PRACTICES

Which numbers did you add first in the first problem? **Explain** why.

Chapter 4

There are different ways to add three numbers.

How can you add 23, 41, and 17?

Think of different ways to choose digits in the ones **column** to add first.

You can make a ten first. Then add the other ones digit. Then add the tens.

$$2\ 3$$
$$4\ 1$$
$$+\ 1\ 7$$

$$3 + 7 = 10$$
$$10 + 1 = 11$$

Add from top to bottom. First add the top two digits in the ones column, then add the next digit. Then add the tens.

$$2\ 3$$
$$4\ 1$$
$$+\ 1\ 7$$

$$3 + 1 = 4$$
$$4 + 7 = 11$$

Share and Show

Add.

1.
```
  33
  34
+ 32
```

2.
```
  47
  21
+  7
```

3.
```
  65
  13
+ 15
```

4.
```
  58
  27
+ 22
```

5.
```
  12
  22
+ 36
```

6.
```
  10
  42
+ 36
```

✓7.
```
  31
  21
+ 16
```

✓8.
```
  30
  29
+ 48
```

Name _____

Add.

9.
```
  22
  27
+ 18
```

10.
```
  26
  31
+ 19
```

11.
```
  24
  11
+ 53
```

12.
```
  33
  43
+  4
```

13.
```
  40
  17
+ 32
```

14.
```
  25
  25
+ 25
```

15.
```
  19
  65
+ 24
```

16.
```
  73
   4
+ 16
```

17. **GO DEEPER** Mrs. Carson is making food for a party. She makes 20 ham sandwiches, 34 turkey sandwiches, and 38 tuna salad sandwiches. How many sandwiches does she make for the party?

_____ sandwiches

18. **THINK SMARTER** Sophia had 44 marbles. She bought 24 more marbles. Then John gave her 35 marbles. How many marbles does Sophia have now?

_____ marbles

Problem Solving • Applications

Solve. Write or draw to explain.

19. **MATHEMATICAL PRACTICE ①** **Evaluate** Mrs. Shaw has 23 red notebooks, 15 blue notebooks, and 27 green notebooks. How many notebooks does she have?

_____ notebooks

20. **MATHEMATICAL PRACTICE ④** **Model Mathematics**
Write a story problem that could be solved using this number sentence.

$$12 + 28 + \blacksquare = 53$$

_____

_____

_____

_____

21. **THINK SMARTER** Mr. Samson gave his students 31 yellow pencils, 27 red pencils, and 25 blue pencils. How many pencils did he give to his students?

_____ pencils

 TAKE HOME ACTIVITY • Ask your child to show you two ways to add 17, 13, and 24.

Name _____

Algebra • Find Sums for 3 Addends

 COMMON CORE STANDARD—2.NBT.B.6
Use place value understanding and properties of operations to add and subtract.

Add.

1.
```
  2 3
  2 0
+ 2 5
```

2.
```
  1 5
  2 2
+ 3 8
```

3.
```
  1 3
  5 2
+ 3 4
```

4.
```
  2 7
  4 0
+ 1 9
```

5.
```
  3 1
  4 5
+ 2 4
```

6.
```
  3 4
  1 1
+ 2 8
```

7.
```
  4 2
  3 6
+ 1 1
```

8.
```
  1 8
  2 2
+ 3 4
```

9.
```
  5 3
  1 9
+ 2 5
```

Problem Solving

Solve. Write or draw to explain.

10. Liam has 24 yellow pencils, 15 red pencils, and 9 blue pencils. How many pencils does he have altogether?

_____ pencils

11. **WRITE** Math Describe how you would find the sum of 24, 36, and 13.

_____

_____

Lesson Check (2.NBT.B.6)

1. What is the sum?

$$
\begin{array}{r}
22 \\
31 \\
+\ 16 \\
\hline
\end{array}
$$

2. What is the sum?

$$
\begin{array}{r}
17 \\
26 \\
+\ 30 \\
\hline
\end{array}
$$

Spiral Review (2.OA.A.1, 2.OA.C.4, 2.NBT.A.3, 2.NBT.B.8)

3. What number is 10 more than 127?

_____

4. Mr. Howard's phone has 4 rows of buttons. There are 3 buttons in each row. How many buttons are on Mr. Howard's phone?

_____ buttons

5. Bob tosses 8 horseshoes. Liz tosses 9 horseshoes. How many horseshoes do they toss?

$8 + 9 =$ _____ horseshoes

6. What number can be written 3 hundreds 1 ten 5 ones?

_____

FOR MORE PRACTICE
GO TO THE
Personal Math Trainer

Name _____

Algebra • Find Sums for 4 Addends

Essential Question What are some ways to add 4 numbers?

Common Core
Number and Operations in
Base Ten—2.NBT.B.6
MATHEMATICAL PRACTICES
MP1, MP6, MP8

Listen and Draw Real World

Show how you solved each problem.

Math Talk MATHEMATICAL PRACTICES 6

Describe how you found the answer to the first problem.

FOR THE TEACHER • Read this problem and have children choose a way to solve it. Shelly counts 16 ants in her ant farm. Pedro counts 22 ants in his farm. Tara counts 14 ants in her farm. How many ants do the 3 children count? Repeat for another problem.

Model and Draw

You can add digits in a column in more than one way. Add the ones first. Then add the tens.

Find a sum that you know. Then add to it.

```
  3  1
  1  4
  2  7    8
+ 2  4
```

THINK:
8 + 1 = 9, then add on 7 more. The sum of the ones is 16 ones.

Add pairs of digits first. Then add these sums.

```
  3  1
  1  4    5
  2  7
+ 2  4   11
```

THINK:
5 + 11 = 16, so there are 16 ones in all.

Share and Show MATH BOARD

Add.

1.
```
   23
   11
   22
 + 31
```

2.
```
   30
   15
    3
 + 25
```

3.
```
   13
   26
   54
 + 12
```

4.
```
   27
    2
   23
 + 13
```

✓5.
```
   45
   14
   35
 + 51
```

✓6.
```
   32
   21
   15
 + 30
```

Name _____

On Your Own

Add.

7.
```
   36
   12
   21
 + 26
```

8.
```
   14
   23
   20
 + 11
```

9.
```
   22
   13
   15
 + 27
```

10.
```
   45
   12
   41
 + 22
```

11.
```
   59
   31
   51
 + 73
```

12.
```
   34
   10
   31
 + 22
```

13. **GO DEEPER** Some friends need 100 bows for a project. Sara brings 12 bows, Angela brings 50 bows, and Nora brings 34 bows. How many more bows do they need?

_____ more bows

Solve. Write or draw to explain.

14. **THINK SMARTER** Laney added four numbers which have a total of 128. She spilled some juice over one number. What is that number?

Math on the Spot

$22 + 43 + $ $ + 30 = 128$

_____

Problem Solving • Applications  WRITE Math

Use the table.
Write or draw to show how you solved the problems.

Shells Collected at the Beach	
Child	Number of Shells
Katie	34
Paul	15
Noah	26
Laura	21

15. **MATHEMATICAL PRACTICE ①** Evaluate How many shells did the four children collect at the beach?

_____ shells

16. **GO DEEPER** Which two children collected more shells at the beach, Katie and Paul, or Noah and Laura?

_____

17. **THINK SMARTER** There were 24 red beads, 31 blue beads, and 8 green beads in a jar. Then Emma put 16 beads into the jar. Write a number sentence to show the number of beads in the jar.

_____

 TAKE HOME ACTIVITY • Have your child explain what he or she learned in this lesson.

Algebra • Find Sums for 4 Addends

Common Core **COMMON CORE STANDARD—2.NBT.B.6**
Use place value understanding and properties of operations to add and subtract.

Add.

1.

```
  1 8
  3 2
  2 3
+   3
```

2.

```
  4 5
  3 1
  2 9
+ 7 2
```

3.

```
  2 4
  6 2
  7 0
+ 3 3
```

4.

```
  8 3
  3 2
  6 1
+ 2 2
```

5.

```
  3 7
  1 5
  3 1
+ 1 2
```

6.

```
  2 1
  1 3
  9 6
+ 1 8
```

Problem Solving

Solve. Show how you solved the problem.

7. Kinza jogs 16 minutes on Monday, 13 minutes on Tuesday, 9 minutes on Wednesday, and 20 minutes on Thursday. What is the total number of minutes she jogged?

_____ minutes

8. **WRITE Math** Describe two different strategies you could use to add 16 + 35 + 24 + 14.

_____

Lesson Check (2.NBT.B.6)

1. What is the sum?

$$
\begin{array}{r}
1\,2 \\
3\,3 \\
5\,6 \\
+\,3\,2 \\
\hline
\end{array}
$$

2. What is the sum?

$$
\begin{array}{r}
4\,1 \\
7\,4 \\
4\,3 \\
+\,2\,0 \\
\hline
\end{array}
$$

Spiral Review (2.OA.A.1, 2.NBT.B.5)

3. Laura had 6 daisies. Then she found 7 more daisies. How many daisies does she have now?

$6 + 7 =$ _____ daisies

4. What is the sum?

$$
\begin{array}{r}
52 \\
+27 \\
\hline
\end{array}
$$

5. Alan has 25 trading cards. He buys 8 more. How many cards does he have now?

$25 + 8 =$ _____ cards

6. Jen saw 13 guinea pigs and 18 gerbils at the pet store. How many pets did she see?

$13 + 18 =$ _____ pets

FOR MORE PRACTICE
GO TO THE
Personal Math Trainer

© Houghton Mifflin Harcourt Publishing Company

308 three hundred eight

Name _____

1. Beth baked 24 carrot muffins. She baked 18 apple muffins. How many muffins did Beth bake?

   Label the bar model. Write a number sentence with a 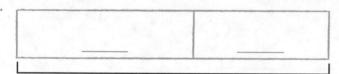 for the missing number. Solve.

_____	_____

   _____

   _____   _____ muffins

2. Carlos has 23 red keys, 36 blue keys, and 44 green keys. How many keys does he have? Circle your answer.

   Carlos has
67
80
103
   keys.

3. Mike sees 17 blue cars and 25 green cars. Choose all the ways you can use to find how many cars he sees. Then solve.

   ○ $\begin{array}{r} 17 \\ +\ 25 \\ \hline \end{array}$ ○ $\begin{array}{r} 25 \\ -\ 17 \\ \hline \end{array}$ ○ $\begin{array}{r} 25 \\ +\ 17 \\ \hline \end{array}$ ○ $\begin{array}{r} 17 \\ +\ 17 \\ \hline \end{array}$

   Mike sees _____ cars.

   Describe how you solved the problem.

   _____

   _____

4. Jerry has 53 pencils in one drawer. He has 27 pencils in another drawer.

Draw a picture or write to explain how to find the number of pencils in both drawers.

Jerry has _____ pencils.

5. THINK SMARTER ✚ Lauren sees 14 birds. Her friend sees 7 birds. How many birds do Lauren and her friend see? Draw quick pictures to solve. Write the sum.

Tens	Ones

_____ birds

Did you regroup to find the answer? Explain.

_____

_____

6. Matt says he can find the sum of 45 + 50 without rewriting it. Explain how you can solve this problem using mental math.

_____

_____

7. Ling sees the three signs at the theater.

Section A	Section B	Section C
35 seats	43 seats	17 seats

Which two sections have 78 seats?

Explain how you made your choices.

_____

_____

_____

8. Leah put 21 white marbles, 31 black marbles, and
7 blue marbles in a bag. Then her sister added
19 yellow marbles.

Write a number sentence to show the number of
marbles in the bag.

_____

9. Nicole made a necklace. She used 13 red beads and
26 blue beads. Show how you can break apart the
addends to find how many beads Nicole used.

$$13 \longrightarrow \underline{\quad} + \underline{\quad}$$

$$+\ 26 \longrightarrow \underline{\quad} + \underline{\quad}$$

$$\underline{\quad}\qquad \underline{\qquad\qquad}$$

$$\underline{\quad} + \underline{\quad} = \underline{\quad}$$

10. **GO DEEPER** Without finding the sums, does the pair of addends have a sum greater than 100? Choose Yes or No.

$51 + 92$	○ Yes	○ No
$42 + 27$	○ Yes	○ No
$82 + 33$	○ Yes	○ No
$62 + 14$	○ Yes	○ No

Explain how you decided which pairs have a sum greater than 100.

_____

_____

_____

11. Leslie finds 24 paper clips in her desk. She finds 8 more paper clips in her pencil box. Choose all the ways you can use to find how many paper clips Leslie has in all.

○ $24 + 8$

○ $24 - 8$

○ $24 + 6 + 2$

12. Mr. O'Brien visited a lighthouse. He climbed 26 stairs. Then he climbed 64 more stairs to the top. How many stairs did he climb at the lighthouse?

_____ stairs